Copernicus

Edward S. Holden

Copernicus

The Astronomer Who Stopped the Sun

LM Publishers

I

Copernicus
Story of the great astronomer

Nicolaus Copernicus was born in Thorn, a town of Prussian Poland, on February 19, 1473. His father, Niklas Koppernigk, was a merchant of Krakau who established himself in Thorn about 1450, and there married Barbara, the daughter of Lucas Watzelrode, a descendant of an old patrician family. The father was chosen alderman in 1465—a testimony of his worth. He had four children: Barbara, who died abbess of the Cistercians at Culm; Katherina, who married a merchant of Krakau; and two sons, Andreas and Nicolaus.

We know little of the childhood of Nicolaus. In 1483 his father died and he was placed in the care of his uncle, another Lucas Watzelrode, who was called to be bishop of Ermeland in 1489, and with whose career that of Copernicus is closely bound up. The boy was educated in Thorn till his nineteenth year, when he was placed in the University of Krakau. The greatest illustration of its faculty was Albertus Blar de Brudzewo (usually written Brudzewski), professor of astronomy and mathematics. The works of Purbach and of Regiomontanus were expounded in his lectures. In the winter semester of 1491-92 Copernicus was matriculated in the faculty of arts, and devoted himself, so it is recorded, with the greatest diligence and success to mathematical and astronomical studies, becoming, at the same time, familiar with the use of astronomical instruments. In the autumn of 1494 Brudzewski left the university, and it is probable that Copernicus did the same. The humanists of the faculty had suffered a defeat at the hands of the scholastics, and the latter now ruled supreme. At Krakau Copernicus studied the theory of perspective, and applied it in painting. Portraits from his hand are praised by his contemporaries.

In the summer of 1496 the youth went to Italy, and in January, 1497, he was inscribed at the University of Bologna, in the 'Album of the German Nation,' as a student of jurisprudence. From 1484 to 1514 the professor of astronomy at Bologna was Dominicus Maria da Novara. He was an observer, a theorist, as well as a free critic of the received doctrines of Ptolemy, although such of his criticisms as we know are not especially happy, it must be confessed. He determined the obliquity of the ecliptic to be 23° 29' by his own observations, which is in error by 1' 20" only, a small quantity for his time. Copernicus was received by him on the footing of a friend and helper, rather than as a pupil; and the association was, without doubt, of great benefit to the younger man. All the systematized knowledge of the time was opened to him; what was known was examined and discussed, not received uncritically. Best of all, observation was practised as a test of theory and as the only basis for its advancement.

The first recorded observation of Copernicus is an occultation of *Aldebaran* by the moon in 1497 at Bologna; in 1500 he observed a conjunction of Saturn with the moon at the same place, and a lunar eclipse at Rome. Other eclipses were observed in 1509, 1511, 1522 and 1523; and positions of Venus, Mars, Jupiter and Saturn in 1512, 1514, 1518, 1520, 1523, 1526, 1527, 1529, 1532, 1537. These recorded observations extend over a period of forty years. Though they are few in number, there is no reason to doubt that they are merely excerpts from a more considerable collection. They were made with very simple wooden instruments constructed by the observer's own hands. One of them, a *triquetum*, was sent as a present to Tycho Brahé in 1584, forty-one years after the death of Copernicus. It was made of pine wood, eight feet long, with two equal cross arms. They were divided, in ink, into 1,000 equal parts, and the long arm into 1,414 parts. This precious relic, together with a portrait of Copernicus, was long preserved in Tycho 's observatory at Uraniborg, and finally removed to Bohemia, where it perished in the confusions incident to the Thirty Years' War (1618-48).

Rheticus once urged upon him the need of making astronomical observations with all imaginable accuracy. Copernicus laughed at his friend for being disturbed about so small an error as a minute of arc, and declared that if he were sure of his observations to ten minutes, he would be as pleased as was Pythagoras when he discovered the properties of the right-angled triangle. Copernicus determined the latitude of Frauenburg to be 54° 19½', which is 2' too small. This seems to us a large error. Even with his instruments he could have been more precise if he had repeated his observations many times. But the determination was excellent for the times, as we may see by remembering that the latitude of Paris was given by Tycho as 48° 10', by Fernel as 48° 40', by Vieta as 48° 49', by Kepler as 48° 39'. His calculated longitude of *Spica Virginis,* which he took as a standard star, was 40' in error. He concluded that Krakau and Frauenburg were on the same meridian—an error of 17½' of arc. The observations of Albategnius, five centuries earlier, were far more precise, and this was not entirely owing to the superiority of the Arab instruments.

At the University of Bologna Copernicus mastered Greek. The knowledge was subsequently utilized in a translation into Latin of the epistles of Theophylactos Simokatta (630 A. D.), which he printed in 1509. This was the only work published by him in his lifetime. The translation is said to be elegant, but the book itself is of comparatively little importance. He had studied it at the university and utilized his knowledge. The book upon which his fame rests—'De Revolutionibus Orbium Cœlestium' —did not appear until the very day of his death, and was published by the care of others. Scipione dal Ferro, the discoverer of the general method of solving the cubic equation, was in residence at Bologna at the same time, and there is little doubt that Copernicus met him also, although there is no record of the meeting. In recording this name we seem to be well out of the middle age. A general solution of the cubic belongs to the modern period, although the Arabs were working on the question in the tenth century.

In 1497 Copernicus was appointed Canon of Frauenburg, which assured to him, for life, an income corresponding to about $2,250 of our money of today, and a leave of absence of three years was granted him to continue his studies in Italy. At a later date he also received a sinecure appointment at Breslau. He had already taken the lesser vows; to the higher he never was dedicated. In 1499 his brother Andreas was likewise consecrated Canon of Frauenburg, and he also matriculated at Bologna (1498) in the faculty of law. Both brothers were represented at home by substitutes, and considerable expense may have attached to this, but it is curious to note that on account of the 'costly living' at the university they needed, and received, remittances from the bishop, their uncle.

In the summer of 1500 his leave of absence expired, and in company with his brother he crossed the Alps to Frauenburg, where both received a new permission to return to Italy. It was stipulated that Nicolaus should study medicine after the completion of his courses in law, in order that he might serve as physician to the Frauenburg chapter. In the autumn of 1501 both brothers were again in Italy, Andreas at Rome, Nicolaus at Padua. The doctor's degree in jurisprudence was conferred upon Nicolaus in 1503, but he remained in Italy till the year 1505 or 1506—nine or ten years in all.

In the archives of Ferrara we read:

1503. Die ultima mensis Maij. Ferrarie in episcopali palatio, sub lodia horti presentibus testibus vocatis et rogatis Spectibili viro domino Joanne Andrea de Lazaris siculo panormito almi Juristarum gymnasii Ferrariensis Magnifico Rectore, Ser Bartholomeo de Silvestris, cive et notario Ferrariensi. Ludovico quondam Baldassaris de Regio cive Ferrariensi et bidello Universitatis Juristarum civitatis Ferrarie, et alijs.

m: Venerabilis, ac doctissimus vir Nicholaus Copernich de Prusia Canonicus Varmensis et Scholasticus ecclesie S. crucis Vratislaviensis: qui studuit Bononie et Padue, fuit approbatus in Jure canonico nemine penitus discrepante, et doctoratus per prefatum dominum Georgium Vicarium antedictum etc.

In the year 1500 Copernicus delivered lectures at Rome before an audience of two thousand hearers, the Archbishop of Mechlin declares. These lectures could not have announced the heliocentric theory, which dates from the year 1506 only, nor could they have been before the university, because Copernicus did not take the degree that admitted him to the privilege of teaching until 1503. He took no degree at Krakau, so far as is known.

Copernicus was now quite free to prosecute his studies in medicine, which he combined with philosophy. The celebrated Pomponazzi was then a member of the faculty, in the prime of his vigor. He had taken his degrees in philosophy and medicine at Padua in 1487, and in the next year, when he was but twenty-six years of age, had been chosen extraordinary professor. It was a custom of those days to choose two professors of each subject in order that their public disputations might stimulate their hearers to independent thinking. The ordinary professor of philosophy was Achillini—a veteran of the strict school of Aristotle.

Pomponazzi remained at Padua until the university was closed in 1509; and in Ferrara till 1512, when he removed to Bologna, where in 1516 he wrote his famous treatise on the 'Immortality of the Soul'—the foundation of his character as a skeptic and of his fame as a philosopher. Into his doctrines it is not necessary to enter at length. Briefly they are that man, standing on the confines of two worlds—the material and spiritual—necessarily partakes of the nature of both. Man is partly mortal (since the human soul depends in some degree on matter) and partly immortal. The soul is, Pomponazzi says, absolutely mortal, relatively immortal. This doctrine was, of course, a denial of the theory of the Roman church. He was vehemently attacked. His book was burned in Venice. Powerful friends among the cardinals protected him in Rome. His university stood by him and confirmed him in his professorial chair for eight years, and increased his salary to 1,200 ducats.

Pomponazzi was a thinker of essentially modern spirit. Reason, he said, was superior to any authority. If, in his teaching of Aristotle, he should find himself in error, "ought I," he says, "to interpret him differently from my real sentiment? If it is said—the hearers are scandalized—well, be it so. They are not obliged to listen to me, or to forbid my teaching. I neither wish to lie, nor to be false to my true conviction." He decides, on psychological grounds, against the immortality of the soul, and then proceeds to build up a system of practical ethics resting on philosophy. Belief is not needed as a basis for ethics—not by cultured men, at any rate. He is the first writer within the christian communion to attempt to establish morality on a foundation of reason. He is a Stoic. "The essential reward of virtue is virtue itself," he says; "the punishment of the vicious is vice, than which nothing can be more wretched and unhappy." Future rewards and punishments are not invoked.

It is worth our while to pause here and reflect that we are hearing a teacher to whom Copernicus listened; to whom all Italy, nay all Europe, attended. This teaching was permitted in Italy. It influenced thousands upon thousands of hearers. Perhaps the tolerant treatment of

Lutherans in Ermeland by Copernicus when administrator of his diocese may have had its origin in ideas received at this time.

There were other men in the faculty with a message for pupils of genius. Aristotle and Plato were expounded from original Greek texts, and the mazy fabrics of the commentators were swept away. Fracastor, who was, by and by, to become an opponent of the heliocentric theory, was a teacher there. He was the first to teach that the obliquity of the ecliptic changed uniformly (1538), in which respect— only—his doctrine was more sound than that of Copernicus. Medicine was expounded by four professors, and dissection of the human body was practised. Marc Antonio della Torre, the instructor of da Vinci, was one of the anatomists. So far as is known, Copernicus did not take his doctor's degree in medicine.

He was, however, skilled in physick, after the fashion of his day, and practised the art during all his life. He was considered, some of his biographers say, 'a second Æsculapius.' We know nothing definite of his medical practise until his later years. From 1529 to 1537 he treated Bishop Ferber, who praises him as the preserver of his life. Duke Albrecht of Prussia called him to Königsberg in 1541 to treat one of his court, and it is of record that the patient recovered.

It does not appear that Copernicus returned to Frauenburg before 1506. He was then thirty-three years of age. All that the world had then to offer in the way of culture was his. He had followed university studies in theology, philosophy, logic, medicine, mathematics and astronomy. He had mastered Greek, and practised painting. He had been the friend or pupil of the greatest teachers of Italy for ten years, and was now established as physician to his uncle in the bishop's palace at Heilsberg, in high station, with an assured income. Up to this period he had shown no original power; but there can be no doubt that he was universally regarded as a man of the highest culture.

His relation to his uncle was that of Achates to Æneas, affectionate and intimate. The bishop of Ermeland was a great noble in a place of power. Affairs of much import to the church had to be treated. The knights of the Teutonic order (founded at Acre in 1190) had conquered the Duchy of Prussia in the thirteenth century. West Prussia had been ceded to Poland in 1466, while East Prussia, including Ermeland, was a Polish fief. A part of the policy of the order was to extend the lordship of their metropolitan Bishop of Riga over the diocese of Ermeland. It was the policy of Bishop Lucas to oppose all such efforts, to attain entire independence, and even to become spiritual over-lord of a part of the territory of the Teutonic order. These plans came to nothing; but a legacy of hatred remained among the knights, who left nothing undone to provoke and degrade the Ermeland bishop and his friends, and to excite disorder in his own territory. The pressure of the invading Tartars on the borders kept the knights occupied, however, and left them little leisure for hostile action. Constant vigilance was required on the part of the bishop, and many journeys to different parts of the bishopric were required.

Copernicus was charged with missions of this sort from the very first. It was during one of these journeys to Petrikau in 1509 that he printed his Latin version of the 'Epistles' of

Theophylactos. Greek epistles—invading Tartars—feudal rights—church privileges—Polish and Prussian politics—these were the preoccupations of his mind. We can hardly think that much time was left for astronomy, yet the lunar eclipse of June 2, 1509, was duly observed. One of Copernicus 's biographers calls him 'a quiet scholarly monk of studious habits—in study and meditation his life passed—he does not appear as having entered into the life of the times.' This is the legend. It is obviously only a small part of the truth. In March, 1512, the bishop of Ermeland died and Copernicus returned to his cloister at Frauenburg. He was now thirty-nine years old.

In the dedication of his 'De Revolutionibus' to the Pope (1542), Copernicus says that it is now 'four nines of years' since the heliocentric theory was conceived. Strictly interpreted this brings the date of its birth to 1506. It is, at all events, safe to say that the idea was elaborated on German, though it may have been born on Italian, soil.

From 1512 to 1516 Copernicus was in constant residence at the Cathedral of Frauenburg, where indeed the greatest part of his life was spent. For two periods (1516-19 and 1520-21) he lived at Allenstein, administering certain estates belonging to his chapter. His observatory was on one of its towers and commanded a wide horizon. Few observations were necessary for his great discovery of the heliocentric motion. He knew beforehand the phenomena to be explained. Ptolemy had offered a solution that had been accepted for fourteen hundred years. Would any other hypothesis explain them? In the first place, Copernicus affirms the rotation of the earth on its axis. The rising and the setting of the stars is caused by this.

The question of the rotation of the earth had been examined by Ptolemy. He rejects the notion, saying: "If the earth turned in twenty-four hours around its axis every point on its surface would be endowed with an immense velocity, and from the rotation a force of projection would arise capable of tearing the most solid buildings from their foundations and of scattering their fragments in the air." The force of projection depends, we know, not only on the absolute velocity of points on the turning earth (and this velocity is immense), but also on the angular velocity about this axis. The latter is slow. The hour hand of a clock turns twice as fast as the earth. The projective force at its maximum is just sufficient to diminish the weight of a ton by six pounds. A feeble force of the sort is not fitted to tear trees up by their roots or buildings from their foundations, as Ptolemy supposed.

Copernicus adopted the theory of a rotating earth, although he was no better able than Ptolemy to explain the difficulty. The science of mechanics was not born till the time of Galileo. The reasoning of Copernicus is: "The rotation of the earth being a natural movement, its effects are very different from those of a violent motion; and the earth, which turns in virtue of its proper nature, is not to be likened to a wheel that is constrained to turn by force." He seeks to escape the difficulty by a trick of scholastic philosophy. No other issue was open in his day. Examples of this sort are well fitted to give us a vivid idea of the state of science in those times. It was not easy for our predecessors to take a forward step. More honor to them that the steps were taken.

In the preface to the 'De Revolutionibus' Copernicus declares that he was dissatisfied with the want of symmetry in the theory of eccentrics and weary of the uncertainty of the mathematical conditions. Searching through the works of the ancients, he found that some of them held that the earth was in motion, not stationary. Philolaus, for example, taught that the earth revolved about a central fire. Copernicus makes no mention of the theory of Aristarchus. We must assume that he did not know it, though his ignorance in this respect is hard to explain. We have no list of his library, which was, however, extensive for the time.

"Then I too," says Copernicus, "began to meditate concerning the motion of the earth; and although it appeared an absurd opinion, yet since I knew that, in earlier times, others had been allowed the privilege of imagining what circles they might choose in order to explain the phenomena, I conceived that I also might take the liberty of trying whether, on the supposition of the earth's motion, it were possible to find better explanations of the revolutions of the celestial orbs than those of ancient times. Having then assumed the motions of the earth that are hereafter explained, by long and laborious observation I found at length that if the motions of the other planets be likened to the revolution of the earth, not only their observed phenomena follow from the suppositions, but also that the several orbs, and the whole system, are so connected in order and magnitude that no one part can be transposed without disturbing the rest and introducing confusion into the whole universe." He looked, he here says, for a new theory because the old one was unsymmetric; and his new theory satisfies because it consistently explains the facts of observation and because it was symmetric. Symmetry of the kind referred to is not essential to a true theory. If any theory explains every fact of observation quantitatively as well as qualitatively, it is to be accepted. Copernicus was not free from hampering presuppositions any more than his predecessors.

"We must admit," he says, "that the celestial motions are circular, or else compounded of several circles, since their inequalities observe a fixed law, and recur in value at certain intervals, which could not be unless they were circular; for the circle alone can make that which has been recur again." In writing this passage his mind was closed to every idea but one. Copernicus knew, far better than most of us, that ovals and ellipses might also serve to represent recurring values, but the thought did not even cross his mind in connection with celestial motions. He was committed to circular motions exclusively, from the outset.

"We are therefore not ashamed to confess," he says, "that the whole of the space within the orbit of the moon, along with the center of the earth, moves around the sun in a year among the other planets; the magnitude of the world (solar system) being so great that the distance of the earth from the sun has no apparent magnitude (is indefinitely small) when compared with the sphere of the fixed stars. . . . All which things, though they be difficult and almost inconceivable, and against the opinion of the majority, we, in the sequel, by God's favor, will make clearer than the sun, at least to those who are not ignorant of mathematics."

The system of Copernicus required thirty-four circles and epicycles—four for the moon, three for the earth, seven for the planet Mercury and five for each of the other planets. Cumbrous as this apparatus appears to us, it was a distinct simplification of the Ptolemaic system as taught in the sixteenth century. Fracastor, writing in 1538, employed sixty-three spheres to explain the celestial motions.

One word must be said of the theory of trepidation which Coper nicus accepted. The precession of the equinoxes was discovered by Hipparchus by comparing his own observations of stars with preceding ones. He saw that the longitudes of the stars changed progressively and fixed the annual change as $1°$ in seventy-five years. Later observers determined the amount of precession by comparing their own observations with preceding ones. The motion of the origin of longitudes—the equinox—is really uniform. An unlucky Jew—Tabit ben Korra—in the ninth century, came to the conclusion that the motion was not uniform, but variable, sometimes at one rate, sometimes at another. The variable motion was the trepidation. Copernicus admitted the reality of this phenomenon and thereby introduced a fault. Tycho Brahe, who had no important data on this point that was inaccessible to Copernicus, rejected the idea of trepidation and freed astronomy from a blemish that had endured for centuries.

It is impossible and unnecessary to exhibit in this place the details of the heliocentric theory of Copernicus. In Kepler's account of Copernican astronomy there is a section on the explanation of the retrogradations of the planets. "Here," he says, "is the triumph of the

Copernican astronomy. The old astronomy can only be silent and admire; the new speaks and gives rational account of every appearance; the old multiplies its epicycles; the new, far simpler, preserves everything by the single motion of the earth around the sun. ' ' In describing the stationary points of the planets he declared: ''Here the old astronomy has naught to say."

We must try to put ourselves in the place of the students of those days who heard the two explanations of the world—the geocentric and the heliocentric—expounded by the same professor in the same lecture-room as alternative hypotheses. Each hypothesis offered a possible explanation. That of Copernicus was so simple that its intellectual acceptance was immediate. It was possible; but was it true? If it were accepted, what implications did it bring in its train? The real difficulty was moral, not intellectual. "Was the whole edifice of Ptolemy to be destroyed? No—some of it was indubitably true. If some, why not all? What was to become of the authority he had held for a thousand years? Was all knowledge to be made over? Even the idea that part of the 'Almagest' was true and part false was not to be lightly accepted.

The conception that every physical problem has one and only one solution was also entirely new; until it was fully received students balanced one explanation against another, and even held two at once, strange as this may seem to us with our new standards in such matters. The heliocentric theory eventually prevailed not because the logic of Ptolemy was broken down, but because all mere authority was weakened.

The dicta of philosophers were looked at in a new light. It was not, in fact, generally received until the day of Newton, though it was sufficiently established by the observations of Galileo and convincingly by the calculations of Kepler. To actually demonstrate the rotation of the earth on its axis we must have recourse to an elaborate experiment like that of Foucault on the pendulum, or to comparisons of the force of gravity in different latitudes; to demonstrate its revolution round the sun it is necessary to measure the time required for light to reach us from the distant planets, or to evaluate the aberration of the light of the fixed stars. It was not easy for the sixteenth century to make a decision. If the heliocentric theory were true, then the planet Venus must show phases like the moon; but no phases could be seen. It required Galileo's telescope to show them. Moreover, the fixed stars must have annual apparent displacements in miniature orbits. None such were visible; none were detected until 1837, when Bessel determined the parallax of a fixed star (61 *Cygni*) for the first time. Galileo sought for them in vain; so did Herschel; so did other astronomers of the eighteenth century with their splendid instruments. The conception of epicycles was retained in the 'De Revolutionibus,' and it seems to us a blemish; to the contemporaries of Copernicus it was a mere analytic device. Newton explains one of the inequalities of the moon's motion by an epicycle, in the 'Principia.'

It is only when we thus consider in detail how the new ideas must have presented themselves to the students of the sixteenth century that we can comprehend the real obstacles in the way of their acceptance. A genius like Kepler could receive them simply on their intellectual merits. Men in general required time to change their point of view, and to accept a novel and essentially disheartening theory. Ptolemy's system of the world was compendious, comfortable, so to say, and easily understanded of the people. Man's central position in the universe flattered his pride and allayed his fears.

Peter the Lombard (1100-60) expresses the accepted view in its baldest form: 'Just as Man is made for the sake of God, that is, that he may serve him, so the Universe is made for the sake of Man, that is, that it may serve him; therefore is Man placed at the middle point of the Universe, that he may both serve and be served.' The new view made man an outcast and placed him in immense and disquieting solitudes. Pascal has phrased the new and anxious fear: 'Le silence éternel de ces espaces infinis m'effraie.'

Astronomers needed accurate tables of the planetary motions in order to predict eclipses and conjunctions. The Alphonsine tables were quite unsatisfactory. The theory of Copernicus was made the basis of new tables—the Prutenic tables—by Reinhold in 1551, and they remained the standard until 1627, when the Rudolphine tables, based on Kepler's theories and Tycho's observations, superseded them. The doctrines of Copernicus were spread by means of almanacs based upon Reinhold's tables rather than by his theoretical works; and they made their way quietly, surely and without any great opposition. Tycho proposed a new (and erroneous) system of the world in 1587. It also had its effect in weakening the authority of Ptolemy. The motions of comets began to be observed with care. It was clear that the doctrine of material crystal spheres would not allow room for their erratic courses. In one way and another the authority of the ancients was broken down and the way prepared for the eventual triumph of the theory of Copernicus.

It is interesting to note the opinions of Englishmen of the sixteenth and seventeenth centuries. Francis Bacon rejected the new doctrines; Gilbert of Colchester, Robert Recorde, Thomas Digges and other Englishmen of the time of Queen Elizabeth, accepted them. Milton seems to hesitate in 'Paradise Lost' (book viii.), which was written after 1640, though he had visited Galileo in Florence in 1638, where, no doubt, Galileo proved the Copernican theory to him by word of mouth. At all events he thoroughly understood it as his description of the earth

> *. . . that spinning sleeps* On her soft axle,
> while she paces even And bears thee soft
> with the smooth air along,

abundantly proves, since in the last line one of the chief objections to the theory is answered.

The heliocentric theory gained powerful auxiliaries in Moestlin, professor of astronomy at Tübingen, and in his pupil Kepler. In 1588 Moestlin printed his 'Epitome,' in which the mobility of the earth is denied; but he accepted the new views probably as early as 1590. Kepler writes: "While I was at Tübingen, attending to Michael Moestlin, I was so delighted with Copernicus, of whom he made great mention in his lectures, that I not only defended his opinions in our disputations of the candidates, but wrote a thesis concerning the first motion which is produced by the revolution of the earth." In 1596 Moestlin, in a published epistle, expressly adhered to the heliocentric theory of the world.

Luther emphatically declared his opinion of the Copernican theory on several occasions. He calls Copernicus 'that fool' who is trying to upset the whole art of astronomy; and refers to Joshua's command that the sun should stand still as a proof that the earth could not possibly be the moving member of the system. Melanchthon, a far more learned man, declared that the authority of scripture was entirely against Copernicus. The attitude of the Roman Church was more indifferent at that time, not more tolerant. Tolerance comes with enlightenment; and both protestant and catholic doctors were, in general, profoundly ignorant of science. When we are thinking of the attitude of the church we must remember that the conflict with Galileo had not arisen. Calvin quotes the first verse of the ninety-third Psalm

—*The World also is established, that it cannot be moved*

and says: 'Who will venture to place the authority of Copernicus above that of the Holy Spirit?'

Such dicta of great theologians are often quoted to demonstrate the existence of an age-long conflict between science and religion. So to interpret them is a sad misconception of the real warfare that has occupied mankind for ages. The veritable conflict has been between ignorance and enlightenment, not in one field only, but in all conceivable spheres.

Before there can be fruitful discussion the 'universe of discourse' must be defined. Things of a like kind can alone be compared. The world of science relates and refers to material things moved by physical forces; and only to these. The world of religion relates and refers only to immaterial things moved by spiritual energies. These worlds are wide apart now. They were widely separated even in the sixteenth century, and they were entirely divided for the highest thinking men even in the middle ages. In either world conflicts are possible. They can only take place between ideas of the same kind; between religion and heresy, or between science and pseudo-science. Theologians decide the issue in one world; men of science in the other. It is the business of philosophers to define and discuss the limits of each world in turn; to determine the validity of conclusions. It is the privilege of poets harmoniously to express imagined analogies between the action of spirit on spirit and of force on matter. It is the dream of seers and prophets to synthesize such analogies into a single system, mingling two universes into one. Whatever may be our hope for the future, the synthesis has not yet been achieved. Theologians have essayed it from one direction, philosophers from another, but the essential distinction remains untouched. There is a world of matter; there is a world of spirit. Men live in both. Their actions are ruled by different and discrepant laws. In the world of spirit the good man is safe and happy, no matter what fate may befall him in the world of physical phenomena. In the latter world no virtue will save the man who transgresses its especial laws. Gravitation, and not goodness, decides whether his falling body suffers harm or is preserved alive.

To Calvin the pronouncement of Copernicus was sheer blasphemy. It seemed to him to lie entirely within the sphere of religion. Judged by the accepted standards of that sphere it was audacious heresy. To Kepler the law of Copernicus lay entirely within the sphere of science. It was to be accepted as true, or rejected as pseudo, science entirely by scientific criteria. Calvin's words fell within one universe of discourse, Kepler's in another. There was no conflict between religion and science as such. Calvin sat as judge of a conflict between religion and a possible heresy. Kepler asked himself if this new assertion was substantial truth or merely error masquerading in a scientific form. Phenomena cannot be judged by criteria belonging to a world to which they are foreign. It is in a light like this that we must examine the relations of such men as Copernicus and Galileo to their times.

The Lateran council (1512-17) appointed a committee to consider the much needed reform of the Church calendar, and in 1514 the help of Copernicus was asked—a proof that he was not only remembered in Rome, but that his reputation had grown since his residence there. He declined to give advice, for the reason that the motions of the sun and moon were, as yet, too imperfectly known. At the request of the chief of the committee, Copernicus continued his researches on the length of the tropical year—a fundamental datum.

In November, 1516, the quiet life of Copernicus at Frauenburg was broken up by his appointment as *Administrator bonorum communium* at Allenstein. The appointment was for one year, but the administration of Copernicus was so successful that he occupied the post during the years 1516-19 and again in 1520-21. His manifold duties in this place brought him again into conflict with the Teutonic knights. The interests of the order and of the church in Ermeland were totally antagonistic. At times open hostilities occurred and towns were besieged, taken and plundered. It is not necessary to follow this harassing strife into the details of Prussian and Polish politics. It is recounted in history as the *Fränkischer Reiterkrieg.* In 1521 Copernicus, then the recognized head of his chapter, was selected to draw up a statement of grievances against the order to be laid before the estates of Prussia. The lands of the chapter of Frauenburg had been overrun, the towns and villages plundered, the peasants had fled or had been killed. The castle of Allenstein, the residence of Copernicus, was itself in danger until it was saved by a four years' truce concluded at Thorn. In such stormy times astronomy was not to be thought of.

It was at this period that Copernicus composed, at the request of the Prussian estates, a memorial on the debasement of the coinage of the country and on the remedies to be adopted. "Money," he says, "is a measure, and like all measures it must be constant in value. What would one say to a yard or a pound whose values could be changed at the will of the measure-makers? The value of money depends not on the stamp it bears, but on the value of the fine metal it contains." Nothing could be clearer than this. His conclusions on the effects of a debased currency on the interests of landlord and tenant are not so sound. Copernicus also proposed to coin all the money of Prussia at a single mint, forbidding the towns to use their ancient privileges, which had been abused. This proposal, as well as others made in the years 1521-30, failed chiefly because Dantzig and other towns were not Mailing to relinquish vested rights. It is interesting to note that in his memorial of 1526 he sets the ratio of gold and silver as 1 to 12.

Bishop Fabian died in 1523. During the ensuing vacancy Copernicus was chosen administrator of the diocese. His duties were harassing. The troops of the order encroached more and more on the church holdings. The Lutheran heresy was also a source of anxiety. The steps taken by the administrator were marked by great tolerance. Before the preaching of the new faith was forbidden outright it was enjoined that it should be refuted by argument. A new bishop, Mauritius Ferber, was chosen in 1523, and a word must be said of the bishop's nephew and coadjutor, Tiedemann Giese. Born in 1480, he became canon of Frauenburg about 1504, and was the intimate and affectionate friend of Copernicus during the whole of his life. It was to Mm that Copernicus confided the manuscript of his great work in 1542. Bishop Ferber died in 1537, and Bishop Dantiscus of Culm was chosen in his place, while Giese by a compromise became bishop of Culm.

The last observation recorded by Copernicus in the 'De Revolutionibus' is dated 1529. From this we may infer that his great work was essentially completed at that time, though it was repeatedly revised afterwards. It had been begun twenty-three years earlier. It was not published until 1543, though its doctrines had been freely communicated to scholars and friends. In 1531 a set of strolling players, set on, it is said, by his enemies among the Teutonic knights and among the Lutherans, gave a little show at Elbing ridiculing the notion that the earth moved round the sun. The play was devised by a certain Dutchman who afterwards became rector of the gymnasium at Elbing. That its satire was understood by the common people proves the opinions of Copernicus to have been fairly well known by his neighbors even at that epoch when absolutely nothing had been printed concerning them. About 1530 a manuscript commentary on the hypotheses of the celestial motions had been prepared by Copernicus for private circulation among men of science in advance of the publication of 'De Revolutionibus.' Two copies of this manuscript still exist, one at Vienna, one at Upsala. At the end of it a resume of his new doctrine is given in seven axioms. (I.) There is only one center to the motions of the heavenly bodies; (II.) this is not the earth about which the moon moves, but (III.) it is the sun; (IV.) the sphere of the fixed stars is indefinitely more distant than the planets; (V.) the diurnal motion of the sun is a consequence of the earth's rotation; (VI.) the annual motion of the sun and (VII.) the motions of the planets are, primarily, not due to their proper motions.

In 1533 Copernicus was sixty years old and applied for a coadjutor. His duties were, at this time, made light for him. In 1532 an observation of Venus is recorded. Other observations were made in 1537. In 1533 he observed the comet of that year. It may be surmised (his memoir on the comet is not extant) that the retrograde motion of this heavenly body confirmed in his mind his criticisms of the system of Ptolemy.

The theory of Copernicus began to be known in Rome, and it was well received. In 1533 Widmanstad, secretary to Pope Clement VII., gave a formal explanation of the heliocentric theory of Copernicus to the pope and to an audience containing several cardinals and bishops. There is no doubt that the theory was received with interest. There is no sign of opposition, and Widmanstad subsequently obtained high honors in the church. The attitude of the Lutherans was, as we have seen, very different. The cardinal-bishop of Capua wrote in 1536 to Copernicus begging him for an explanation of his system.

In 1537 Dantiscus became bishop of Ermeland. All the canons of Frauenburg, Copernicus included, supported his nomination. Copernicus was known, however, to be a warm friend of Giese, who should have succeeded, as coadjutor, to his uncle's bishopric, but who was elected to that of Culm by a compromise. Difficulties soon arose between Copernicus and his new bishop, and the breach was widened in various ways. The bishop, himself a man of loose morals, ordered Copernicus to send away his housekeeper, on the assumption of illicit relations between the two, and kept the accusation alive by various official letters. Bishop Dantiscus oppressed Copernicus in various ways and remained his enemy in spite of certain advances on the part of the latter. If Copernicus ever feared the persecution of the church on account of his scientific teaching—of which there is little evidence—it was because his bishop stood ready to use every and any weapon against him.

Copernicus gained an ardent disciple in George Joachim of Rhaetia, known to us as Rheticus. He was born in 1514 and made his studies at Nuremberg under Schoner to such effect that he was appointed to be professor of mathematics at the University of Wittenberg in 1537, at the age of twenty-three. In May, 1539, he visited the great astronomer of Frauenburg chiefly to study his doctrines of trigonometry, and his trigonometric tables. Copernicus was then sixty-six years of age and his enthusiastic and loyal guest was twenty-five. He was received cordially and at once set himself to study the manuscripts of Copernicus. His visit extended itself from a few weeks to more than two years, and he became a firm believer in the new heliocentric astronomy, which he was well prepared to receive and to expound.

A letter from Rheticus, written a few months after his arrival at Frauenburg, affords one of the very few personal views of Copernicus that have come down to us. The letter was published with a long Latin title, in 1540, and is known as 'Narratio Prima.' "I beg you to have this opinion concerning that learned man, my preceptor: that he had been an ardent admirer and follower of Ptolemy; but when he was compelled by phenomena and demonstration, he thought he did well to aim at the same mark at which Ptolemy had aimed, though with a bow and shafts of very different material from his. We must recollect what Ptolemy has said: 'He who is to follow philosophy must be a freeman in mind.'" "My preceptor was very far from rejecting the opinions of ancient philosophers from love of novelty, and except for weighty reasons and irresistible facts. His years, his gravity of character, his excellent learning, his magnanimity and nobleness of spirit are very far from any such temper (of disrespect to the ancients)." This letter, addressed by Rheticus to his old master Schöner, was the first easily accessible account of the new theory. The life-giving sun, he says, is placed in its appropriate place, and a single motion of the earth explains all the planetary motions. All is harmony as if they were bound together with a golden chain. He praises the great simplicity and reasonableness of the new doctrine, as well as the almost divine insight and the uncommon diligence of the master. He had formerly no idea, he says, of the immense labor required in such works, and the example of Copernicus leaves him in astonishment. Copernicus had made a complete collection of all known astronomical observations, and by these his theory was tested. The master was not content until every hypothesis had been fully proved.

Rheticus showed his admiration for Copernicus not only in these public, but also in private, ways. Books that he presented to the master (which are often annotated by Copernicus 's own hand) are still to be found in various libraries of Sweden, where they were taken after the plundering of Ermeland in the thirty years' war. At Wittenburg Rheticus and his colleague Reinhold, Copernicans both, were by the conditions of their professorships obliged to teach the Ptolemaic system, just as Galileo, at Padua, a Copernican, had to confine himself to the exposition of Sacrobosco. It may safely be surmised, however, that their pupils did not leave them without hearing something of the true doctrines. In the 'Narratio,' Rheticus, who was a firm believer in astrology, uses the data of the 'De Revolutionibus' as bases for wide-reaching astrological predictions. They are of no interest in themselves, but as the letter was written under the eye of Copernicus, they lead to the conclusion that they were not disapproved by the latter. So far as I know, this is the only evidence for the belief of Copernicus in astrology. We have no horoscopes from his hand but, like all his contemporaries, he probably gave it a place among the sciences.

Rheticus deserves the gratitude of all calculators for his table of trigonometric functions (sines, tangents, secants) to ten decimal places, for every 10" of the quadrant, published in a huge volume by his pupil, Otho, under the title 'Opus Palatinum de Triangulis.' The tables of Rheticus are the basis upon which Vlacq founded his great tables, and they have served as models for many followers. Lansberg's tables appeared fifteen years after the 'Opus Palatinum' and lightened the immense labors of Kepler.

Toward the end of the year 1541 Rheticus returned to Wittenberg carrying with him a part of Copernicus 's manuscript—a treatise on 'Trigonometry'—which he printed in 1542. The complete manuscript of the 'De Revolutionibus' was sent by Copernicus to his old friend Giese, the bishop of Culm, for such disposition as he thought best. The bishop sent it to Rheticus to arrange for its printing at Nuremberg, and to see it through the press. It fell out that the printing had to be confided to Andreas Osiander, a Lutheran minister interested in astronomy. The book was published early in 1543, and a copy reached Copernicus on May 24, the very day of his death.

Osiander prefixed to the volume an introductory note which he did not sign, as follows:

Scholars will be surprised by the novelty of the hypothesis proposed in this book, which supposes the earth to be in motion about the sun, itself fixed. But if they will look closer they will see that the author is in no wise to be blamed. The aim of astronomy is to observe the heavenly bodies and to discover the laws of their motions; the veritable causes of the motions it is impossible to assign. It is consequently permissible to imagine causes, arbitrarily, under the sole condition that they should represent, geometrically, the state of the heavens, and it is not necessary that such hypotheses should be true, or even probable. It is sufficient that they should furnish positions that agree with observations. If astronomy admits principles, it is not for the purpose of affirming their truth, but to give a certain basis for calculation.

The best authorities affirm that Osiander 's apology, which he had suggested to Copernicus as early as 1540, was unauthorized.

Osiander made many changes in the text also, and added the last two words of the title under which the book was printed—' De Revolutionibus Orbium Cœlestium. ' Readers of our day universally interpret the apology to be an attempt to forestall theological opposition and persecution. They remember the conflict of Galileo with the church. But Osiander was a protestant divine, Copernicus a catholic priest. It is passing strange to conceive that a Lutheran schismatic should intervene to shield an orthodox catholic from accusations of heresy. Moreover, Copernicus had good reasons for believing that the princes of the church would receive his work favorably. His doctrine had been known to them since 1530. He knew, however, that several powerful university teachers —Fracastor for one—opposed it. Ought we not to interpret the apology as an address to men of science? Whewell justly remarks that

Copernicus seems to consider the opposition of divines as a 'less formidable danger' than that of astronomers. It is difficult to admit that Osiander dared to prefix this note without the authorization of Copernicus, or, at least, of Rheticus. There seems to be no reason to doubt that it was addressed solely to men of science.

The words of the apology represent the exact point of view of the ancients, and are entirely opposed to the attitude of modern science. Centuries of experience have taught the modern world that there is one and only one solution to a scientific problem. Modern science is a search for such unique solutions. Anything less definite is an hypothesis to be held tentatively and temporarily, it may be even alternatively with another, or others. The theories of the Greek philosophers were, in general, held by them primarily as hypotheses. Their whole attitude towards scientific certainty was thus entirely different from our own. In the time of Copernicus the minds of most men were cast in the ancient temper. It is, in fact, from his century that the new insight dates. This is not to say that colossal geniuses like Archimedes or Roger Bacon did not work in what we call the modern spirit. It is simply to confirm that most of the contemporaries of Copernicus belonged, in this respect, to the ancient world. The apology expressed exactly their attitude. The attitude and temper of the modern world are entirely different; they are perfectly formulated in these words of Pascal (in french): "*Ce n'est pas le décret de Rome sur le mouvement de la terre qui prouvera qu'elle demeure en repos; et, si l'on avait des observations constantes qui prouvassent que c'est elle qui tourne, tous les hommes ensembles ne l'empêcheraient pas de tourner, et ne s'empêcheraient pas de tourner avec elle."*

It required this very book of Copernicus to suggest the pregnant phrase of Pascal.

In the letter of dedication to the Pope—Paul III.—Copernicus speaks in his own name. His words are simple and serious, full of dignity and conviction:

I dedicate my book to your Holiness in order that both learned men and the ignorant may see that I do not shrink from judgment and examination. If perchance there be vain babblers who, knowing nothing of mathematics, yet assume the right of judging on account of some place of Scripture perversely twisted to their purpose, and who blame and attack my undertaking, I heed them not and look upon their judgments as rash and contemptible.

He is here referring to divines. The following is addressed to astronomers. *Though I know that the thoughts of a philosopher do not depend on the judgment of the multitude, his study being to seek out truth in all things so far as is permitted by God to human reason, yet when I considered how absurd my doctrine would appear I long hesitated whether I should publish my book, or whether it were not better to follow the example of the Pythagoreans and others who delivered their doctrine only by tradition, and to friends.*

The doctrine of Copernicus was first formally judged by the Roman Church in 1615 when Galileo was before the Inquisition in Rome. The judgment was in these terms:

The first proposition, that the sun is the center and does not revolve about the earth, is foolish, absurd, false in theology, and heretical, because expressly contrary to Holy Scripture.

The second proposition, that the earth revolves about the sun and is not the center, is absurd, false in philosophy and, from a theological point of view at least, opposed to the true faith.

In the year 1616 the works of Copernicus were placed upon the Index 'until they should be corrected,' and 'all writings which affirm the motion of the Earth' were condemned at the same time. The congregation issued a notice to its readers in 1620, thus conceived:

Although the writings of Copernicus, the illustrious astronomer, on the revolutions of the world have been declared completely condemnable by the Fathers of the Sacred Congregation of the Index, for the reason that he is not content to announce hypothetically certain principles concerning the situation and motion of the earth, which principles are entirely contrary to the sacred Scripture, and to its true and Catholic interpretation (which can absolutely not be tolerated in a Christian man) but dares to present them as indeed true; nevertheless, because this book contains things very useful to the republic, it has been unanimously agreed that the works of Copernicus ought to be authorized, so far printed, as they previously have been authorized, correcting, however, according to the following notes, the passages in which he does not express himself hypothetically, but affirmatatively maintains the motion of the earth; but those which, in future, will be printed must not be so printed save with the following corrections, which are to be placed before the preface of Copernicus.

The corrections follow; they are not numerous or important.

The works of Copernicus were still on the Index in the year 1819. In the following year Pope Pius VII. approved a decree of the Congregation of the Holy Office that the Copernican system, as established, might be taught, and in 1822 'the printing and publication of works treating of the motion of the earth and the stability of the sun, in accordance with the general opinion of modern astronomers, is permitted at Home.' Centuries before this date the real question had been judged; but its formal settlement in the Roman Church was postponed to our own day.

The judgments of the Congregation of the Index upon the heliocentric theory were an incident in the history of the relations of Galileo with the authorities at Rome, and they can best be understood in connection with that history. Something, however, may be said of them here. It is to be observed that the first proposition is condemned because it is contrary to scripture, heretical, false in theology, *absurd* and *foolish*; and the second because, from a theological point of view it is opposed to the true faith, *false in philosophy and absurd*. The words not in italics relate to judgments upon points of doctrine. The words in italics relate to judgments upon matters of philosophy or of science.

It was entirely competent for the Congregation of the Index to render decisions upon matters of theology which were binding upon all catholics. The committee was organized and existed for that purpose. Every institution, religious or secular, must decide for itself on matters of the sort. Not to do so is sheer suicide. The competence of the Roman church and of the Congregation of the Index to decide *for* itself *questions of what is opposed to its faith, contrary to scripture,* false in theology, is not to be denied. This was a conflict of theology with an alleged heresy. Copernicus was a member of the Roman Church. The soundness of his theological opinion was a matter for doctors of theology to settle in their own church in their own way. They did not decide it, however, until they had taken the advice of astronomers who pronounced the heliocentric theory to be baseless. (Delambre, 'Astronomie moderne,' i., p. 681.) Tycho Brahe, also— a great authority—had declared it to be 'absurd and contrary to the scriptures.' These two points are often forgotten by writers of the Martyr-of-Science School.

On the other hand, no one can admit for a moment the right or the competence of the Congregation or of the Church to pronounce final judgment upon a question of philosophy or of science. The whole world is now agreed that it is an impertinence for a body of theologians to pronounce upon a question of science, precisely as it would be for a congress of scientific men to pronounce upon a point of theology.

The reasons that led the Congregation of the Index to take this fatal step must be considered in connection with the history of Galileo. It will not be out of place here, however, to attempt to understand the mistaken point of view of the churchmen responsible for the decision.

For fourteen hundred years the theory of Ptolemy had ruled. In 1543 Copernicus proposed a new and revolutionary system. In its essential point the system was true, as we know now; we also know that it was false in asserting that the planets moved in circular orbits (they really move in ellipses), in accepting trepidation as an incident to precession, and in other matters of the sort. It even asserted, falsely, that the center of the orbit of the earth and not the sun was the center of planetary motion, so that in a strict sense it was not even a heliocentric theory. The theory of Copernicus was not proved to be true, in its essential feature, until Galileo discovered the phases of Venus, in 1610. Is it any wonder that doctors of the church five years afterwards were not convinced? They were profoundly ignorant of science and not in the least interested in science as such. Any one of them could recollect that Tycho Brahe, the greatest astronomer of his time, had in 1587 made a theory of the world which placed the earth at its center. He, then, did not agree with the theory of Copernicus. He expressly rejected it. It could easily be recollected, also, that in 1597 Kepler had proposed his first theory of the world, in which the planets were arranged according to fanciful and false analogies with the shapes of the five regular solids of Plato. It is now known that the systems of Tycho and of Kepler were both false. Ought the church doctors to have accepted them when they were proposed? In 1609 Kepler proposed a second theory of the world based on elliptic and heliocentric motion. How could the doctors know that this second system was the true one, as indeed it was? Kepler was still alive. How could they know that he would not propose a third theory? They had seen the doctrine of Ptolemy denied by Copernicus; the doctrine of Copernicus denied by Tycho; the doctrine of Tycho denied by Kepler's first system; the doctrine of Kepler's first replaced by that of his second system. All this had occurred within their own memories. In scientific theories as such they had no interest whatever; they were solely concerned for religion. Is it surprising that they did not promptly accept a theory which they did not understand?

It was, however, a profound and inexcusable error for them to condemn it; and by so doing they, unwittingly, dealt a heavy blow to the church. For once, theology engaged in a warfare with science; and the issue was an overwhelming and deserved victory for science. There have not been many such conflicts. Very exceptional conditions are required to bring them about, as may be seen in the long history of Galileo.

It is very difficult to form a vivid conception of the whole character of Copernicus either from his works or from his portraits. We know far too little of his history and too little of the time in which he lived. I have found no summary in any of his biographies that can be called satisfying and I have never been able to make one for myself. I venture to reprint that of Bertrand, and to enclose in parentheses those parts that we positively know to need modification or correction.

Capernic est pour nous tout entier dans son livre. Sa vie intime est mal connu. Ce qu'on en sait donne l'idée d'un homme ferme, mais prudent, et d'un caractère parfaitement droit; tout entier à ses spéculations et comme recuelli en lui-même; il aimait la paix, la solitude, et le silence. Simplement et sincèrement pieux, il ne comprit jamais que la verité pût mettre la foi en péril, et se reserva toujours le droit de la chercher et d'y croire. Aucune passion ne troubla sa vie; (ou ne lui connaît même pas de commerce affectueux et intime); ennemi des discours inutiles, il ne rechercha ni les éloges ni le bruit de la gloire; indépendant sans orgueil, content de son sort et content de lui-même, il fut grand sans éclat, et, ne se révélant qu'a petit nombre de disciples choisis, il a accompli une revolution dans la science (sans que, se son vivant, l'Europe en ait rien su).

The system of Copernicus belongs to him alone. It is not the system of Philolaus or of Aristarchus . . . but his own. His name is justly attached to it on account of the care with which he explained its every part, brought out all its phenomena, discovered the causes of these precessional movements which had been known for eighteen hundred years, and explained only by the hypothetical existence of an eighth sphere which made a revolution in 36,000 years around the axis of the ecliptic, while, at the same time, it was constrained to turn daily about the axis of the equator to account for the rising and setting of the stars. It is then Copernicus who really introduced the motion of the earth into astronomy, not merely into academic disputations; it is he who demonstrated how the revolution of the earth about the sun explained the succession of the seasons and the precession of the equinoxes; it is he who showed how simply the retrogradations of the planets are explained by the unequal velocity with which they traverse their concentric orbits about the sun; it is he who put astronomy on new foundations and who opened the way for all later researches. It is to Kepler's enthusiasm over the new truths that we owe the discovery of the true shape of the planetary orbits, and the laws of their motion. The idea of the motion of the earth was unfruitful among the ancients because it was never entertained with seriousness. Its adoption by Copernicus is the beginning of modern astronomy. (Delambre).

The mountain peaks that cluster closely round the Lick Observatory in California are of different heights and were unnamed when the corps of observing astronomers took possession of the newly established station. Names were assigned to them in the order of their heights — Copernicus, Galileo, Kepler, Tycho and Ptolemy. One of the staff of observers, who greatly distinguished himself during his short career at the observatory, objected to the assignment of the name of Copernicus to the highest peak. Copernicus was, no doubt, a great astronomer, he said, but was he preeminent? Should not the highest peak have been assigned to another? The objection is answered the moment the relation of Copernicus to the whole thought of the world is comprehended. His skill as a mere observer, his power as a mere geometer, is not in question. His place is not to be assigned by narrow criteria like these. What was the attitude of man towards everything not himself before the day of Copernicus? towards things divine, things spiritual, things natural? What is his view of the world now? The changes are so fundamental, extensive and bewildering as not to be described, much less estimated, except by a long series of separate steps, each one opening new worlds in religion, philosophy, science, art, technics. To name them all would be to summarize the entire history of human progress for three hundred and fifty years. In the long stairway of ascent Copernicus established the foundation stone. Tycho, Kepler, Galileo, Newton, Kant, Laplace,

Herschel, Darwin (to speak only of men of science) each laid successive steps upon it. Until the first was firmly laid no building, no advance, was possible. We stand to-day in a high place of vantage won for us by the master builders of more than three centuries. Without Copernicus their work would have been in vain. The modern world is erected upon foundations that he laid.

NIKOLAUS COPERNICUS.

II

Synthetic History of Copernicus[1]

Modern astronomy may be said to have begun with Copernicus. Previous to his time the received theories of the structure and motions of the universe were incorrect, inconsistent, and incomprehensible, and did not explain the inexact observations that were referred to them. He gave to science a correct theory, in which exact observations have found clear and consistent explanations.

Copernicus lived at the time of the awakening of knowledge, and was a part of it. The idea that the earth moved around the sun was not new; it had been uttered before, but, like many other thoughts that had been

---

[1] From Sketch published in Popular Science of June, 1891.

expressed among the ancients and then slumbered through the middle ages, it, being contrary to the received notions, was frowned on by authority and was refused a hearing. Copernicus saw, what an intelligent observer could not fail to see, that none of the systems then known could account for the motions of the stars. He had met the most distinguished astronomers of his own time. He was acquainted with all the systems of the ancients; and the more he examined them the more he was astonished at the want of harmony and inconsistency that marked them. "I then took pains," he says, "to read again all the books of philosophy that I could get, to assure myself whether I could find any different opinions from those which were taught in the schools concerning the motions of the spheres of the world. And I saw first in Cicero that Nicetas had expressed the opinion that the earth moves.

Then I found in Plutarch that others had had the same idea... . Further, the leading Pythagoreans, Archytas of Tarentum, Heraclides of Pontus, Echrecrates, etc., taught the same doctrine, according to which the earth is not motionless in the center of the world, but turns in a circle, and is far from holding the first rank among the heavenly bodies." Pythagoras had learned the same doctrine; Timæus of Locris was very precise in announcing it, when he called the five planets the "organs of time on account of their revolutions," and added that we should have to suppose that the earth was not immovable in the same place, but that it turned around itself and was also carried along in space. Plutarch says that Plato, who had always taught that the sun turned around the earth, changed his opinion toward the end of his life, and regretted that he had not placed the sun at the center of the world, the only place that became it. Three centuries before Christ Aristarchus of Samos, according to Archimedes, composed a work, now lost, defending the doctrine of the movement of the earth against the opinions of philosophers to the contrary, in which he said that "the sun continues immovable and the earth moves around the sun, describing a circular course of which that star occupies the center." Passing to the Romans, this system of Aristarchus was modified into one like that of Tycho Brahe.

In his review of the ancient systems, Copernicus was most drawn, according to M. Biot, "to that of the Egyptians, which made Mercury and Venus revolve round the sun, and put Mars, Jupiter, Saturn, and the sun in motion round the earth; and to that of Apollonius of Perga, which made the sun the common center of all the planetary motions, while the sun itself revolved around the earth—an arrangement that became the system of Tycho Brahe. Copernicus was impressed with these systems because he found that they represented well the limited excursions of Mars and Venus around the sun, explaining their movements, direct, stationary, and retrograde, an advantage which the system of Apollonius extended to the superior planets. The astronomical planets were thus no longer simple sports of the imagination to him. He had studied them experimentally, and had found the conditions which they must satisfy. The hardest part of his discovery was made. On the other hand, he perceived that the Pythagoreans had taken away the earth from the center of the world and put the sun there. It seemed to him that Apollonius's system would be simpler and more symmetrical if it was modified in this sense, so as to suppose the sun fixed in the center, and the earth revolving round it. He had seen also that Nicetas, Heraclides, and other philosophers, while they placed the earth in the center of the world, had ventured to give it a movement of rotation upon itself, producing the phenomena of the rising and setting of the stars and the alternations of day and night. He still more approved the theory of Philolaus, who, taking the earth away from the center of the world, had given it a rotation on its axis and another motion of annual revolution around the sun. And, although it might seem difficult and even absurd to take the earth from the center and make a simple planet of it, yet, as other astronomers before him had taken the liberty of imagining circles in the sky to explain phenomena, he thought he might be permitted to look for some other arrangement, with a moving earth, which would establish a more simple order in the motions of the stars. Thus, taking what is true from each system and rejecting all in them that was false and complicated, he composed that admirable whole which we call the system of Copernicus, and which is really only the correct arrangement of the planetary system to which we belong." "After long researches" Copernicus himself said, "I am convinced that if we refer the—motions of—the other planets to the revolution of the earth, calculation will agree well with observation. . . . I do not doubt that mathematicians will be of my opinion, if they will take the pains to make themselves acquainted,—not superficially but profoundly, with the demonstration which I shall present in this book."

He reasoned that "every displacement manifest to our view proceeds either from the object perceived or from the subject which perceives, or from the unequal motions of the two, for an equal and simultaneous motion of the object and the subject could cause no semblance of displacement. The earth is the place whence the movement of the sky is presented to our view. Every motion starting from the earth is reflected in the sky, which will appear to move in the opposite direction. Such is the diurnal revolution, which appears to involve the whole universe except the earth. If now we suppose that-the sky has none of this motion, but that the earth turns around itself from west to east (in a contrary direction from the apparent motion of the sky), we shall find that it is really-so." Among the chief arguments in support of this view, the astronomer insisted especially on the immensity of the sky—as compared with the size of the earth: "The whole mass of the earth," he said, "vanishes before the grandeur of the sky; the horizon divides the celestial sphere into halves, which could not be if the earth bore any proportion to the extent of the sky, or if its distance from the center—of—the—universe was perceptible! Compared to the sky, the earth is only a point; it is as a finite quantity compared with an infinite quantity. It is no more admissible to suppose the earth resting in the center of the universe. What! to believe that immensity turns every twenty-four hours around an insignificancy!" So the inequalities in the movements of the planets their forward and backward—movements and stationary positions were referred to two causes: the movement of translation of the earth and the proper motions of the planets; correctly, as modern astronomers explain them, only Copernicus was not able to give details and exact figures?

Ptolemy had argued against the idea of these motions of the earth, because if the earth were translated through space it would leave all the loose things on it behind; and, if it turned on its axis from west to east, it would be impossible for bodies to make any headway to the eastward, for, whatever the rate of their motion, the earth would always reach a given point in that direction first. Hence the former idea was the most ridiculous of all (πάντων γελοιότατα) and the latter altogether ridiculous (πάνυ γλοΙόταον). These arguments seemed unanswerable, and had been received, with Ptolemy's theory, till they had become almost an article of faith. It required a courage which we can only weakly comprehend at this day for a student to fly in the face of the world, of science and religion, and take the solar system to pieces, to put it together again, and to say, after all, that it is the earth which moves and not the sun. Copernicus was slow in venturing before the public with his theory. He began the formulation of his system in 1507; but he wisely determined to make thorough work of the matter, and publish nothing that he could not support with carefully considered argument and evidence. He would not be satisfied with reconciling general appearances with his theory; he would go into details and show how it fitted individual phenomena. He would show how all the movements of the heavenly bodies could be accounted for and predicted by it; even how those phenomena which had hitherto proved unaccountable, the stationary positions and retrograde motions of the planets, and the precession of the equinoxes, found explanation in it. In the mean time reports had got into circulation respecting his new theory, and the public wanted to know what it was. Astronomers were waiting for it, and he was urged to publish it. But he delayed, revising his sheets daily for the insertion of corrected data, and adding new results; and he shrank from the inevitable conflict with the prejudices of the day. These prejudices were already beginning to make their mark. Men of science could accept his views or give them utterance, so far as they had been made acquainted with them, but the general public was against them. He was ridiculed in a comedy; but his gravity and self-restraint carried him safely through all these trials. At last he permitted his friends to publish the work, which he dedicated, in deprecation of clerical censure, to Pope Paul III, in order, as he said in the dedication, that no one should accuse him of running away from the judgment of enlightened men, and that the authority of his Holiness, if he should approve the work, might secure him against the stings of calumny. "I believe," he also said, "that as soon as what I have written in this book concerning the motions of the earth is known, a cry of shame will be raised against me. I am, further, not so much in love with my ideas as to be careless of what others might think about them. And, although the thoughts of the philosopher differ from the aims of the crowd, because he proposes to seek for the truth, so far as God has given it to human wisdom to do, I am not yet ready to reject entirely opinions which seem to be at variance with mine... . All these motives, together with the fear of becoming—on account of novelty and apparent absurdity—an object of ridicule, had nearly caused me to renounce the enterprise. But some friends—among them Cardinal Schomberg and Tidium Gisius, Bishop of Kulm—succeeded in overcoming my repugnance. The last, especially, insisted most earnestly on my publishing this book, which I had kept on the shelf, not nine years, but nearly thirty-six."

The book (*De Revolutionibus Orbium Cælestium*) was printed at Nuremberg, under the care of Rheticus, one of Copernicus's pupils, in 1543. Although Copernicus had till that time been enjoying excellent health, he had then been attacked by a dysentery; and this had passed into a paralysis, with loss of his mental faculties, when the first copy of the book was given to him only a few hours before his death. He saw it and handled it, but was too far gone to exhibit any signs of appreciation of it, or for his friends to be able to know how he was affected by it, or whether he realized what it was. The first edition of the *De Revolutionibus,* which is now very rare, was followed by a second edition in 1566, and a third in 1617. Seventy-three years after the death of its author, on the 5th of March, 1616, it was condemned by the Congregation of the Index "for containing ideas set forth as true on the positions and motions of the earth entirely contrary to the Holy Scripture."

The first work recording the labors of the astronomer was the letter published by Rheticus under the title *Ad Clar. V. de Jo. Schonerum de Libris Revolutionum eruditiss, Viri et Mathematici excellentiss. Rev. Doctoris Nicolai Copernici Torunnœi, Canonici Varmiensis, per quemdam juvenem Mathematicœ studiosum, Narratio prima,* Dantzic, 1540; reprinted, with a eulogium, at Basle, 1541. The works of Copernicus are *De Revolutionibus Orbium Cœlestium Libri VI,* Nuremberg, 1543; reprinted at Basle in 1566, with the letter of Rheticus, and also included in the Astronomia Instaurataol Nicolas Muller, Amsterdam, 1617 and 1640; a treatise on Trigonometry, with tables of sines, entitled *De Lateribus et Angulis Triangulorum,* Wittenberg; *Theophylacti*

Scholastici Simocattæ Epistolæ morales, rurales, et amatoriæ, cum Versione Latina. There are also the treatise on money, already mentioned, and several manuscript treatises in the library of the bishopric of Wiarmia.

The tomb of Copernicus, which was exactly like those of the other canons of Frauenburg, was adorned with a Latin epitaph by the Polish Bishop Cromer, in 1581. It was repaired by Napoleon I in 1807, and so placed that it could be seen from all parts of the church. A statue of Copernicus by Thorwaldsen was erected by subscriptions from the Polish people, in 1829, in the Casimir Palace at Warsaw. The Polish clergy, invited to attend the ceremonies, refused, because his. book had been condemned by the Holy Office in 1616. Another monument to him, by Tieck, was erected at Thorn in 1853.